Creating Your Own Perfume With A 1700 Percent Markup!

Third Edition

by Philip Goutell

ii

Contents

Part Two:

Working with professionals to create your fragrance 27

Illustrations

Introduction

The original text for *Creating Your Own Perfume With A 1700 Percent Markup!* was very simple because what we did was very simple. Yet we created an original *brand* and 1,000 bottles of fragrance at a cost of less than $1.50 per bottle. We sold every one of those 1,000 bottles for $26.95. You can calculate the markup yourself.

Part One of this book details the sources and methods we used to create *our* fragrance. Today the cost of our components has changed very little and now, thanks to the internet, it is much easier to find the supplies you need. After explaining how we did it, I go through the same steps as to how *you* can do it today, with a lot more resources at your command than we ever had.

Part Two of this book is a guide for those investing enough money in a fragrance to contract with professionals for the services they need. Both **Part One** and **Part Two** detail the same steps but

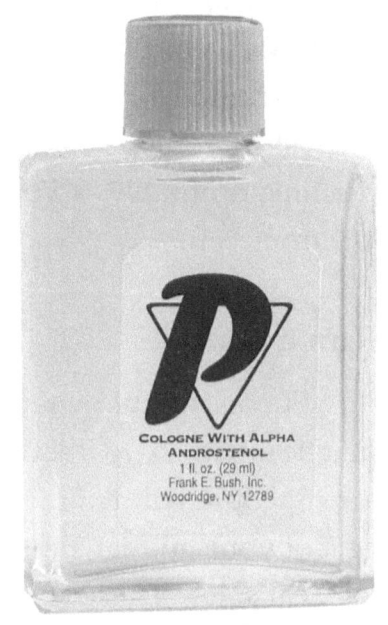

This is "P," the men's cologne we created with a 1700% markup. After a market test with a dummy product, 1,000 bottles were produced at a cost of slightly less than $1.50 per bottle. All 1,000 bottles were sold by us at $26.95 plus $3.50 postage & handling. We sold directly to consumers. Marketing expenses were minimal. The project proved very profitable.

Part One focuses on inexpensive workarounds for the hands-on developer. **Part Two** is focused on decisions that confront the entrepreneur when working with professionals.

Do you have a marketing opportunity?

Before spending money to produce a fragrance be clear about how you will sell it! Are people ready to buy your perfume? Can you *count* them ... and count *on* them? Do you have the tools you will need to *communicate* with them? Where will they go to *buy* your perfume? Will you ship it to them? Will they pay your price? You can't sell perfume to people who are indifferent to you and have *zero* interest in your perfume. To make money with perfume — to sell perfume profitably — you need *friends*, with *money*, who are eager to *buy* from you.

Run a test

Successful marketers test. You should test too if you can find a practical way to do it. Look for tricks you can use on a small scale to see if people will really buy your fragrance. Take, perhaps, ten percent of your budget and put it into a real market test aimed at making real sales. Then, based on your results, project the results you might expect from a rollout. If your test flops, put the brakes on your project. Find out what went wrong and make corrections. Tune it up and get it right before you spend more than you can afford to lose.

Run the numbers

Suppose your fragrance will retail at $35 and your wholesale price will average $17.50. Now suppose you will put $5,000 into this project. If your cost per bottle is $3.50, your margin will be $14.00 so you'll need 357 (5,000/14.00 = 357) sales to get your money back. But if your cost per bottle is $7.50, that number will rise to 500 (5000/10 = 500). It makes a difference.

Running numbers, even estimates, before you finalize your fragrance and packaging will help you appreciate the importance of watching every penny spent toward the cost of your fragrance. The greater your cost per bottle, the more units you have to sell just to get your money back. Your goal, of course, is to sell many times more units than this break even number.

Avoid "The Ten Million Dollar Launch" syndrome

While all fragrances go through essentially the same steps in their development and marketing, the larger, proven brands have a smoother, well oiled path from initial concept to retail distribution and sales. The notes that follow explain "how it works" for established companies in the market — but not (yet) for you!

Major brands are constantly launching new fragrances, sometimes under their own corporate names; sometimes in the name of a celebrity with whom they have a licensing and profit sharing agreement. Budgets for creation, production, and marketing for an individual fragrance may easily reach the multi-million dollar mark. A successful launch might bring back the investment in a year or two but, to be a real hit, the fragrance must sell successfully for *years*.

Promotion

Before the fragrance is created, plans are developed for its promotion. From these plans an idea emerges of what the fragrance should be. This is expressed in a written brief which is then submitted to several giant fragrance creation houses.

Developing the scent

Each fragrance creation house that receives the brief will develop, at their own expense, a fragrance based on the specifications given to them in the brief. Samples will be submitted to the marketer. The marketer selects one sample as the winner. The company that created it now gets the contract to produce the fragrance, both for the perfume and for associated products. The formula for the fragrance remains the property of the fragrance creation house which retains the exclusive right to produce it, but it can only be produced for the marketer that commissioned it.

The bottle

An original bottle will be designed and produced, usually in conjunction with the overall packaging design for the new fragrance. Once the design has been finalized and a quote from a bottle manufacturer accepted, bottles may be produced by the hundreds of thousands, and perhaps millions.

The cap (overshell) and spray

While the bottle will most likely make use of a standard spray pump, the overshell for the pump will almost certainly be of an original design as part of the overall look the design team has given the packaging. This too may be produced by the hundreds of thousands, or millions.

The box and packaging

The box will be a custom shape designed to fit the bottle and continue the design team's graphic theme. Depending on the retail price point for the fragrance, the box may make use of a specialty paperboard and, possibly, a protective corrugated insert. Again, these will be produced by the hundreds of thousands, or millions.

Filling and assembly

To put the finished product together, components will typically be delivered to a contract packaging house for filling and assembly. The fragrance oil — the "juice" — will be blended with alcohol and water, the bottles will be filled and capped with their spray pumps and overshells, boxed, cellophane wrapped, and packed into master cartons for shipping to stores or warehouses designated by the marketer.

Filling may take place in modest batches to meet the requirements of the marketer's current orders. Typically a larger stock of all components will be on hand at the filling house so that assembly can quickly take place "as needed."

The lesson

If you are an investor or part of an investment group seeking to launch a fragrance for the first time, what has been described here as "The Ten Million Dollar Launch" is *not* how you will go about launching your fragrance.

More is going on here than just the money. The global fragrance marketers — Coty, Estée Lauder, L'Oreal, and a handful of others — have *relationships*. Major chain stores are ready to stock their latest fragrances, major advertising agencies are ready to help promote them, major fragrance houses are ready to help them develop and produce new fragrances. Bottle makers work with them to overcome technical hurdles. Contract packaging houses are ready to clear the decks for them, shoving lesser clients aside to give them immediate attention, even if it delays *your* job.

None of these vendors will do this for you.

This does not mean you can't launch your own perfume profitably. It does mean that you will be asked to pay for services that would be free to a major marketer. It does mean you cannot expect to get the most favorable credit terms, even if you have good credit. It does mean that your job may be assigned a lower priority by vendors. And it does mean you will benefit by working with an experienced guide who knows and has worked with the vendors you will be using.

Now for your project ...

Whether your plan is to produce your new fragrance by hand, yourself, or to work with professionals to develop your fragrance, you will find both **Part One** and **Part Two** helpful. But always keep in mind, it is *your market* and *your* abilty to sell that will make it profitable.

— Philip Goutell

June, 2018

Part One

Producing your own perfume or cologne yourself

The original edition of this book was a detailed description of how we "created [our] own perfume with a 1700 percent markup." The perfume was a cologne which we simply named "P."

(1) Producing "P" cologne

Before making our commitment to "P" cologne we ran a small, inexpensive test with a dummy fragrance. Results were profitable, even though our dummy product cost us a good deal more than what "P" would ultimately cost.

We launched "P" knowing that we needed to sell just 60 bottles to break even. Our marketing test had shown that selling 60 bottles was a near certainty.

Our original plan called for a filling house to fill and cap our bottles. When they informed us our job was too small for them and they wouldn't do it, we already had orders coming in.

Either we could refund these prepaid orders ... *or we could get busy and produce the fragrance ourselves!*

Our components

Look at a bottle of perfume. What do you see? Nothing more

than a bottle filled with liquid scent, sealed with a cap or spray and labeled. The bottle goes inside a box and the box may be sealed with shrink wrap to suggest that it contains the genuine fragrance and has not been tampered with.

Thanks to the nature of our business at the time, we felt we could dispense with a box. We would be shipping orders in small, padded, self-seal mailers. Their small size and light weight would give us a low postal rate. Had we boxed our fragrance and continued to use self-seal mailers, the box would have been crushed in shipping and look messy on arrival. Improved packaging would have greatly increased our shipping cost. So we saved the cost of a box.

We purchased bottles with sprinkler necks. This eliminated the

Two bottles side by side. The bottle on the left has a full opening at the neck and will be closed with a spray pump. The bottler on the right has a "sprinkler neck" and needs only a simple cap to close it. Note how the opening in the neck of the bottle on the right is constricted so that only a limited amount of fragrance will be splashed out when in use. Note also that due to this constricted neck, you can't fit a spray pump to a sprinkler neck bottle.

need for spray pumps. Instead we could seal the bottles by hand with inexpensive, screw-on plastic caps. More money was saved; more markup was achieved.

Our fragrance

Our fragrance came from Novarome in New Jersey. Between the time of our first order and our second, Novarome was purchased by Robertet, a major French fragrance business with headquarters in Grasse and offices around the world.

Novarome was very helpful but the first step was for us to tell them what we wanted in a fragrance. I now had to evaluate every men's fragrance I could sample at fragrance counters of local stores. I purchased a number of fragrances to test and settled on the scent of Giorgio's *Red for Men*. Novarome assured us they could supply a good match.

Please understand, we were not trying to knock off Giorgio's fragrance. Our plan was to promote a brand of our own — "P" — with no reference to Giorgio's *Red for Men* which, I was certain, few if any of our customers had ever smelled.

Novarome sent us a box of samples ... of their *Red for Women*. The samples were a hit with the women in our office but we had to ask Novarome to try again. The next box contained the men's samples.

We received six small bottles. There were two versions of the "Red" scent we had designated and each of the versions had been blended with three different quantities of alcohol. A price was quoted for each fragrance oil alone and then for each version of the fragrance mixed with the different levels of alcohol. The price list looked like this:

Men's Fragrance 101 $48/lb
@10% in Alc 85 $40.00/gal
@15% in Alc 85 $56.50/gal
@20% in Alc 85 $73.50/gal

Men's Fragrance 102 $22.50/lb
@10% in Alc 85 $22.00/gal
@15% in Alc 85 $30.00/gal
@20% in Alc 85 $37.50/gal

Notice how we were buying perfume by the gallon and selling it by the ounce. Also notice that the quotes for the fragrance oil itself unmixed with alcohol were by weight, pounds in this case but today quotes in kilograms are more common. Fragrance oils are generally sold by weight.

Alcohol, on the other hand, is sold by volume. When you're producing a perfume yourself and your fragrance vendor does not offer to mix their fragrance with alcohol for you, not only will you need to find a vendor who can supply the alcohol, you'll need to calculate how many pounds or kilograms of fragrance oil you'll need to give you the volume (gallons or liters) needed to fill your bottles.

You can do this by knowing the volume (liters, gallons) that a particular weight (kilograms, pounds) of the fragrance oil will yield. Generally your fragrance vendor will be able to give you this information or simply do the calculation for you once you specify the volume you need.

The percentages given in the price quotes refer to the ratio of fragrance oil to alcohol. "@10%" refers to a mixture of 10% fragrance oil and 90% alcohol; "@15%" refers to a mixture of 15% oil to 85% alcohol; "@20%" refers to a mixture of 20% oil to 80% alcohol.

As to "Alc 85," the designation of the alcohol itself, the "85" refers to a mixture of 85% pure alcohol and 15% (usually de-ionized) water. Using the proof system this would be referred to as 170 proof alcohol.

Various alcohol to water ratios can be used in perfumery but smaller companies such as ours generally settle for what a vendor offers.

Discussion

At one time "cologne" referred to a different concept in fragrance than "perfume." It was more of a health tonic and often would be drunk. Today *cologne* simply indicates a men's fragrance while fragrances for women are designated *perfume*. The technology behind the creation of the fragrance oil is the same for both.

Women's fragrances (perfume) are generally marketed in sub-categories with designations such as "EDT" (eau de toilette), "EDC" (eau de cologne), "EDP" (eau de parfum), etc. but these designations simply refer to the level of alcohol and water blended with the fragrance oil. Typically less alcohol and water means a higher price. Occasionally a fragrance may be sold with 100% alcohol or even no alcohol at all. But while these variations may be more expensive and thus more "exclusive," consumer preferences generally favor a moderate amount of alcohol and water blended with the oil.

The alcohol, in the United States and many other countries, will be "denatured," meaning a small amount of something has been added to it that, while not making a noticeable change to its scent characteristics, will make it unsuitable for beverage use. Thus it escapes high taxation.

The U.S. government provides formulas by which alcohol can be denatured for use in perfume. One commonly encountered formula is "SDA 40B":

§21.76 Formula No. 40-B.

(a) Formula. To every 100 gallons of alcohol add:
One-sixteenth avoirdupois ounce of denatonium benzoate, N.F., and 1/8 gallon of tert-butyl alcohol.
(b) Authorized uses. (1) As a solvent:
052. Inks.
111. Hair and scalp preparations.
112. Bay rum.
113. Lotions and creams (hand, face, and body).
114. Deodorants (body).
121. Perfumes and perfume tinctures.
122. Toilet waters and colognes.
141. Shampoos.
142. Soaps and bath preparations.
210. External pharmaceuticals, not U.S.P. or N.F.
410. Disinfectants, insecticides, fungicides, and other biocides.
450. Cleaning solutions (including household detergents).
470. Theater sprays, incense, and room deodorants.
485. Miscellaneous solutions.

In the United States, to purchase more than a few gallons of denatured alcohol a year a special license is needed. In my experience, by the time you're ready to purchase enough alcohol to need the license, you're ready to take your job to a filling house that has an alcohol license and can provide all the alcohol you need.

As to the amount of alcohol used for any perfume (or cologne!), lighter fragrances use more alcohol; heavier fragrances use less. Alcohol evaporates faster than fragrance oil so less alcohol and more oil gives the fragrance a bit longer life, but the choice of aroma materials used to *create* the fragrance make a much larger contribution to this effect called "persistence".

As to the ratios cited in our price quote above, 20% is a very heavy amount of oil for a men's fragrance. Typically a men's fragrance would be over 90% alcohol and water. Yet we went with the more expensive formula with the greatest amount of oil, the $73.50 a gallon choice. In our ignorance (at that time) we simply decided that "more expensive" meant "better," which is not necessarily the case, but in this case our selection proved popular with our customers.

Our plan was to bottle this fragrance in 1 fluid ounce (29 ml) bottles. There are 160 fluid ounces to the gallon so, to fill 1,000 bottles we would need 6.25 gallons (1,000 / 160 = 6.25). At $73.50/gallon, our fragrance would only cost $459.38. To be on the safe side, knowing there could be some waste in the bottling process, we placed an initial order for 8 gallons. The total cost now came to $588.00 plus $21.50 for UPS shipping. Even if we were to have used all eight gallons to fill our 1,000 bottles, our cost for the fragrance would only be $0.61 per bottle. As waste was not a large problem, our actual cost per bottle for fragrance was closer to $0.52.

The original order from Novarome arrived in two 5 gallon carboys. Later when we reordered our order came in a single drum.

Bottles and caps

The bottle we purchased was a "Chanel No. 5 type" with a sprinkler neck. At the time, selecting a sprinkler neck added a bit to the cost but it allowed us to go without a spray which saved us about two dollars per bottle. (Purchased in larger quantities, the spray might have cost well under a dollar.)

Our bottles were purchased as surplus from the W. Braun Company which was moving away from supplying bottles to companies like ours and today no longer exists. I cannot recall our original cost per bottle but when we needed a new supply of the same bottles — Italian glass from Baralan International, 1-ounce size —

our cost was $363 per thousand from Arrowpak International, less than $0.37 each. They were packed 288 to the case and this new order was for seven cases.

There can be a considerable difference in pricing between what an authorized distributor may quote for a small order and what a closeout house might charge for the same bottles, if they have them. The drawback in buying a closeout is that you might not be able to get more of the same should you need them. The savings, however, can be very tempting and often quite practical, and the quality will be the same.

Caps can be an issue as they should fit your bottles perfectly, not just "close enough." Our first caps (black) were fine and cost about $0.10 each which, today, seems to me like a high price. Later when we needed more caps, we could only find one in white plastic which was not a perfect fit. While we would make them work, the threads were not a perfect match for the threads on the necks of the bottles and about one out of four or five cracked when we forced them on. We had to watch for small, undetected cracks once the caps were in place as in time those bottles with cracked caps would begin to leak.

While I would have preferred more black caps, white was all we could find at the time and we really didn't know how to scour the market for other sources. The white caps were priced at $87.40 per thousand and we had to purchase a box of 5,000 caps. I still have some of these white caps left over from this project as, in time, I found a new source of (correct) black caps.

Looking back, I was pleased with the bottles and less than pleased with the white caps but there was much we needed to learn about tracking down sources and negotiating with vendors.

Labels

At the time, while we had experience with labeling, we didn't have experience with labels themselves. We had other products that were ordered 10,000 or 20,000 at a time and for these, the labeling was silk screened onto the bottle. We simply provided the artwork for the job, the legally required text and our company graphics.

Labels were ordered after we learned we would be assembling our fragrance ourselves, by hand. This was fortunate because it allowed us to purchase self-stick labels which would be neat and easy to apply. In order to keep the price down, we purchased our labels in a stock size. Getting the exact size we wanted would have involved a major additional expense for the die cutting and the special print run. Silk screening was ruled out as the cost of silk screening such a small number of bottles would have raised our cost per bottle dramatically.

As to label pricing, black ink on the glossy white label was the least costly option and, for this particular project, black and white was perfectly suitable. The text on our labels was simply the "P" name and the minimal legally required information: ingredients (SD alcohol, water, fragrance), source (our company), and address (our company's city and state address).

Filling the bottles

When we placed our order for sprinkler neck bottles we hadn't considered how we would fill them. The plan had been to have them filled by the outside contractor but when that plan got canceled we found ourselves looking at two 5 gallon carboys of fragrance and 1,000 empty, 1-ounce bottles with a constricted hole in the neck through which we would have to pour — or inject — the fragrance.

We needed a simple, efficient, and rapid technique by which we could fill these bottles by hand without wasting too much of the fragrance (at $73.50 a gallon!)

We tried several methods. We knew we first had to pour the fragrance from the jugs into some kind of smaller container. I found that a Pyrex measuring cup worked well and also allowed us to monitor our usage. From the Pyrex measuring cup we tried using eye droppers to make the fill but the bulb was too small. Using an eye dropper, the work of filling each bottle was so painfully slow that with 1,000 bottles to fill, it would take forever. Thanks to this filling problem, by the end of the first day we had made no progress at all.

The next morning our office manager arrived with a turkey baster — the stainless steel kind that ends in a small, screw-on extension that looks like an over-sized hypodermic needle. The extension easily fit through the sprinkler neck and we were home free.

Here is a turkey baster, right from the store. This proved to be our instrument for filling perfume bottles. Shown below, metal turkey baster with rubber bulb at right and removable metal tip at left. The tip and bulb come apart for cleaning. The metal tip fits nicely into sprinkler neck bottles.

I've been using a turkey baster to fill perfume

bottles ever since, even ones with an open neck. It was a good solution.

When filling bottles by hand, the only caution is not to fill beyond the shoulder of the bottle, the point where the bottle begins to narrow into the neck. It is important to leave this space unfilled to allow for expansion of the fragrance.

If you fill the bottle all the way to the top leaving no air space at all, you have the potential for heat expanding the fragrance, building up pressure inside the bottle, and in some cases building up enough pressure to blow the cap off the bottle. Should this happen (and it once happened to me!), your customer will not thank you.

Always leave air space at the top of the bottle. Don't fill your bottles beyond the shoulder. Leaving that space empty does not cheat your customer, it protects your customer from what can be a big mess.

Finishing the job

Once the bottles were filled, we simply screwed on the caps to seal them. Before applying the label, the last step in our production process, we wiped each bottle with a dry towel to remove any stray fragrance that might have spilled a bit in the filling process, although there was very little of this, but we wanted the glass to be clean and dry when we applied the labels so they would stick properly.

If we had a box, at this point we would have boxed the fragrance. Boxes would have arrived fully printed but lying flat. The boxes would easily be squeezed open, the bottles inserted, and the tops of the boxes closed. Shrink wrapping would have been the final touch but it can be difficult to do a neat job of this and doing it one box at a time by hand is slow business.

The results

All 1,000 bottles were sold at $26.95 each plus $3.50 for shipping and handling. Additional bottles were produced and sold at the same price. We never discounted the product. I no longer have sales records for our "P" cologne but I'd guess that sales for this, our first fragrance, ran around $30,000, all thanks to a few hundred dollars invested in a test and then an initial investment of about $2,000. It was a very gratifying project.

(2) Setting up your project

It is quite practical to produce 1,000 bottles of perfume or cologne on your desktop or a good steady table. Also, you can have all your supplies on hand but bottle your fragrance in batches as it is needed for your retailers. This is how large companies do it.

Your fragrance

Unless you are a perfumer creating your own unique fragrance from scratch, your fragrance will be an existing oil, available to anyone. Your first step is to decide what smell you want. Unless you plan to go to the expense of hiring a perfumer to create an original scent, you will find yourself studying fragrances already on the market and picking one you think could be right for your project. It is easy to find a commercial source of best seller smell-likes.

While the scent you select will not be unique — others could be buying the same product from the same source — when you sell it it will *become* unique because of the *brand* you put on it. This brand name will be unique and it is through this brand name that you will promote your fragrance.

Unless you plan to promote your fragrance as a knock off, a less expensive replica of a famous brand, it is wise to select a scent that will NOT immediately be recognized as a version of a famous brand. The more likeable but less recognizable your scent is, the more that scent becomes associated with *your* company and the name *you* have given it.

If you have an offbeat idea for a fragrance you can take that idea to a perfumer to see if he or she can make it for you. But unless you have a very strong aesthetic sense and can guide, and afford, the services of a really talented perfumer, the outcome may just seem like another version of an existing scent.

Once you locate a source for a fragrance oil that is acceptable to you, be sure you are dealing with a true commercial source, a company that sells in *bulk* rather than a consumer retailer who typically sells a few ounces at a time.

Your alcohol

Alcohol can pose the greatest obstacle to your project. You need alcohol to blend with your fragrance but, in the United States and elsewhere, even perfumers alcohol is regulated by the government and, while you can purchase a modest quantity from online sources, these sources are limited in what they can sell you per calendar year, and what they are allowed to sell you isn't enough for 1,000 bottles of perfume.

For our "P" cologne, both Novaoome and Robertet supplied the fragrance already mixed with alcohol so when we were starting out with our first project, alcohol was not a problem for us. For our second fragrance we bought oil by the pound from a different vendor who did not supply the alcohol. They did, however, introduce us to a filling house with an alcohol license and the filling house blended the oil and alcohol for us.

In the U.S., getting your own alcohol license involves filling out paperwork and paying an annual fee. Generally when you get to the point where you need more than four gallons of alcohol a year it is more convenient to establish a relationship with a filling house and have them do the blending. While you're at it you can ask what they would charge to fill and cap your bottles.

If your plans call for producing no more than 400 or 500 bottles of fragrance annually you should be able to just buy the alcohol you need over the internet. If you give it some thought you might find ways to get a little bit past the annual limitations.

Blending your fragrance

Once you blend your alcohol and fragrance oil, you'll want to allow this mixture to age for about 30 days. Why? Because it takes time for the mixture to fully blend. During that 30 day period, if you were to sample your fragrance periodically you would find that, as the days pass, the scent changes; it gets richer; nicer. It costs nothing to let your fragrance sit for a while and you are making your fragrance better.

You need one or more containers for blending. If you use more than one container, say several glass jugs, it is essential that you mix the exact same ratio of oil to alcohol in each jug. If a filling house mixes your oil and alcohol they will likely return it to you in a single drum. Hopefully the drum will be of a plastic suitable for the storage of fragrance. Metal drums, even when lined, can pose a problem with oxidation, particularly around the bung. Oxidation destroys perfume. It will make it completely unsellable. I tell you this from personal experience.

If you receive your job back from a filling house in a steel drum and are not planning to bottle for more than 90 days it would be smart to transfer your finished fragrance to either a proper plastic drum or, if the quantity is no more than ten gallons, to glass jugs.

If you are doing your own blending and have managed to purchase enough alcohol for your job, and your alcohol is pure "100 proof," you can add water to it to lower the proof and make it "nicer" for your perfume or cologne.

De-ionized water is typically mixed with the alcohol. It can be purchased through chemical supply houses. You are likely to find the shipping charge greater than the cost of the water. Some perfume makers use distilled water which can be found in your local supermarket.

Preferences vary as to the best water to alcohol ratio but they mostly fall in the 5 to 15 percent water to reduce the alcohol to 190 to 170 proof. Blending simply involves pouring the water and alcohol into a container in the correct proportions, shaking it a bit, and letting it sit for 24 hours. For my own fragrances I mostly use 10 percent water to 90 percent pure, 200 proof, SD-40B alcohol.

If your alcohol is less than 200 proof you do not have to add water as water has already been added. (That's why it is less than 200 proof!)

Mixing your oil with alcohol

Pour your oil in first. Add the alcohol slowly letting the mixture blend. How much oil? How much alcohol? Another decision you have to make. There are no hard and fast rules.

Older guides give set ratios for EDT, EDC, EDP, etc. but today these guides are not strictly observed. Marketers do what they need to do to please their clients ... while making money.

For a men's fragrance, 5 to about 7-1/2 percent oil to 95 to 92-1/2 percent alcohol generally works well but for my own men's fragrances I have used up to 20 percent oil to 80 percent alcohol.

For women's fragrances, ratios might go from about 7-1/2 to 20 percent oil to 92-1/2 to 80 percent alcohol. More alcohol gives you a lighter fragrance. More oil gives you a heavier, longer lasting fragrance. In North America, lighter fragrances are generally favored. Elsewhere, heavier fragrances are popular.

Your bottles and closures

The more bottles you are buying, the more choices you will have for both bottles and closures. If you are planning to produce only 50 or 100 bottles and you want them with a spray, you will have to hunt a bit and even then you may find only one or two choices.

Bottle with screw-on spray pump. "Top Hat" overshell is shown at the left. Another pump is shown between the overshell and bottle. Note the height of both pump and overshell. If you are doing hand assembly, it's hard to get around the use of screw-on sprays like this. Alternatives can only be fitted to a bottle by machine.

If you need hundreds or perhaps thousands, you will have

many more choices. But even here you are limited by your need to find a bottle to which you can attach the spray pump by hand (unless the capping will be done by a filling house) as you'll need a screw-on (continuous thread) pump in the same finish (neck size) as your bottles. It is certainly worth searching through websites and catalogs for every bottle and pump that might work for you but in

be broken. Press-on sprays are difficult to find in the U.S. (this one came from the U.K.) and would not be suitable for mass production without a proper capping machine. Note the length of the plastic dip tube. Before the pump is mounted, the dip tube will be trimmed with scissors to just reach the bottom of the bottle.

Here's a bottle that takes a press-on fine mist spray pump. The spray can be mounted to the bottle without machinery but it's a tricky job and some bottles and pumps will likely

the end it is likely you will find your practical choices to be limited. You may come across some elegant solutions and then find they are not priced for bulk ordering and thus will be too expensive for you.

If you are producing a men's fragrance and are planning to use a sprinkler neck bottle, you will again be amazed at how few choices you have.

Spend time researching the possibilities. And think forward. Will you be developing more than one fragrance? If so, it could pay to buy your bottles in bulk, so you can be sure of having the same style bottle and closure for all your fragrances, giving them a family look. Sometimes in buying a closeout and getting more bottles than you need you still save money — and you are now ready for your next project.

A possible alternative to a screw-on spray is the snap-on pump. These can be attached (to matching bottles) by hand or by machine but they are generally sold without matching overshells (to cover the spray's actuator, preventing accidental sprays) and attaching these pumps to your bottles by hand is a bit of a slog, something you would not enjoy if you had to attach hundreds of them yourself.

The important point to watch for in purchasing your bottles is the need for a matching closure, spray pump or cap. Your safest bet can be to get both components from the same vendor. Even then you will want to run your own tests to make sure there won't be any leaks.

Filling your bottles

Large, automatic filling machines work with drums of fragrance. Hoses run from the drum to the filling machine. These machines are suitable for filling thousands of bottles a day.

Smaller, hand operated filling machines exist. Their cost can run around $2,500 so they aren't for one time use but rather for someone developing their own line of fragrances.

All of these filling machines, automatic and manual, are simply larger versions of the metal turkey basters we have used to fill bottles. Like the turkey baster, the tip can fit into a sprinkler neck as well as a regular neck bottle.

I continue to recommend a turkey baster for filling bottles on your own. The technique is simple. Pour some of your fragrance into a container such as a 4-cup (1,000 ml) glass kitchen measuring cup. Fill your turkey baster from it. For a 1-ounce bottle, two squirts will generally get you up to the shoulder of the bottle — the point at which it begins to narrow to the neck — and STOP at this point. You must leave air space between the shoulder and the pump or cap for expansion. If you fill your bottle up to the very top you will have a very big problem on a hot day!

Boxing your fragrance

The only way you can box your fragrance properly is to pay for a custom box. This involves working with a box manufacturer who will create a die based on the specifications you supply: length, width, and depth. A minimum order is generally over 1,000 pieces and, more commonly, 5,000.

So boxing your fragrance will require some thought. If you are producing 1,000 or more bottles, the economics aren't so bad. If you are producing less than 1,000 bottles, you have to give some thought to how badly you need a box and how you might develop your presentation without a custom box.

Websites offering packaging solutions for gift shops and small boutiques are a source for products that can dress up your bottles without your having to go to the expense of a custom box. Supplies you can find include decorative bags and non-fitted boxes that can be stuffed with shredded paper or other fill to make the bottle fit snugly

When your fragrance is on display in a retail store, the "look" of your package is important. But to keep your numbers in line, you must consider ways you can get a clean, professional look with materials that fall within your budget.

Shrink wrapping your box

A small shrink wrap machine is not expensive. You need the machine to hold rolls of shrink wrap and cut it (a heated wire does this). Then you need a heat gun to shrink the wrap. It sounds easy but it can be difficult in practice.

If you box your perfume, shrink wrapping the box can make it look very professional, if you do a good job of shrink wrapping. If not, shrink wrapping will take away from the look and can even make it a disaster, looking like someone re-packaged a returned fragrance. This is something to avoid.

Personally I have never been good at shrink wrapping boxes, but you may have the gift.

Part Two

Working With Professionals To Create Your Fragrance

If you have a good marketing plan and a bit of money you almost certainly will not want to be filling bottles with a turkey baster or sticking labels on thousands of bottles by hand. In your role of fragrance promoter you will contract with professions to produce your fragrance under your direction. The steps will be the same as if you were doing it all by hand, yourself, but others will now be doing the hands-on work and at least some of the tasks will be automated. Before going farther, let's review what you have to do.

(1) Overview of your project

Producing a perfume or cologne involves gathering up the required components and then putting them together in a finished product. For the mechanical steps of purchasing and production, competent professional guidance is available. For the creative and artistic side of your perfume venture, all will be on you.

The elements

Buy a fragrance and examine what you have just purchased. There is a box wrapped in cellophane. On the box in addition to the name of the fragrance you will find the name and address of the marketer and a list of ingredients. You will also find a barcode for inventory tracking purposes.

Inside the box is a glass bottle, labeled and filled with fragrance. The bottle is typically sealed with a spray pump which is covered by a cap or overshell to add a decorative touch and prevent accidental sprays. (In some cases, men's fragrances in particular, a screw-on cap might be used to seal the bottle.) The labeling on the bottle might be one or more stickers or it could be printing on the bottle itself. To produce your own perfume or cologne you simply purchase and assemble these components.

Finding a qualified guide

To complete your project successfully you will have to interact with a number of vendors who will provide the components and services you need. Think of those who are chosen as your team.

But how do you choose them? How do you get the team members to work together? You need a team manager, a person who can help you decide which players to bring on board and which players to avoid. You need players who can work together well, for your benefit. Your first step is to find this team manager-guide.

There are consultants who undertake this role and can guide you through the required purchases and final assembly. If you find a good one, one you feel comfortable working with, work with that person.

If you don't happen to have a connection with this kind of consultant, my suggestion is to find help through a contract packager who regularly produces cosmetics AND fragrance. We call this kind of company a FILLING HOUSE because it is in the business of assembling all the components needed to produce a perfume or cologne and then filling the bottles, packaging them, and sending them on their way to you, ready to sell.

The filling house doesn't supply the individual components. That will be up to you. But they can guide you to reliable sources and explain to you what you will need from each of these vendors. All of your purchases will be shipped to your filling house for filling and assembly.

Because they assemble these components for a number of accounts, the people at the filling house know which vendors can be relied on to make a job go smoothly. If you can first find the right filling house they can introduce you to qualified vendors for the components you need and, because you now have an introduction to these vendors through the filling house, the vendors are more likely to take your requests seriously and guide you through the technology of their particular components.

An alternative plan NOT recommended

In shopping for your components you will come across a significant number of vendors and see some lovely possibilities and prices. There can be a temptation to enter into relationships with various vendors without considering how well they and their products will work together with products and services supplied by others.

You may get away with it but you might also find yourself with components that don't quite fit together — mismatched bottles, spray pumps, and overshells for example — or fragrance oil not quite right for your alcohol so it turns cloudy — or an entire job held up because one all-star vendor works to his own timetable rather than yours. A group of all-stars does not always make an effective team and, should problems arise, who will take the blame? No one! The problem will be on you!

Barcodes

Most retailers of any size will expect to receive your fragrance barcoded with a UPC (Universal Product Code). This means you will have to register a barcode for your fragrance so it can be printed on your boxes. A barcode is good for exactly one version of ONE product so if you are offering multiple versions of your fragrance, each version must carry its own barcode.

The barcode on the package — either a 12-digit Universal Product Code (UPC) in the United States or a 13-digit EAN (originally called "European Article Number" but now sometimes known as a "International Article Number") — is a unique, universal identifier of the product to which it is attached. Each variation of the same item, a different size bottle for example, requires a code of its own.

Barcodes are assigned by an international non-profit association called "GS1." In the United States the organization is "GS1 US." To receive your barcode directly from GS1, you pay a one time enrollment fee and then an annual maintenance fee.

Some retailers may allow you to use codes obtained from third-party vendors. These codes are issued by a company which has its own GS1 registration so a code you receive will reference their company name rather than yours. The code, however, will be universally valid and permanent.

A UPC or EAN itself carries no information other than identifying the GS1 registered company. All information about the item that gets keyed into a retailer's inventory control system will come from you.

If you have a product code for your fragrance at the time you are preparing the graphics for your packaging, the product code can be incorporated into the design. If you are going to begin making sales without a UPC but anticipate that in time you may need one, leave a space on your box where a sticker can be applied at a later date without obscuring the essential text or graphics on the package.

Asian sources

Almost all the information in this book is based on North American and European sources. Today Asia is exploding with opportunities for perfume development and marketing. For those with the ability to negotiate across borders and are willing to deal with international shipping, Asian sources can provide all that is needed, often at costs considerably lower than Western sources. Connections can be made through trade shows such as EastPack and WestPack and online through Alibaba.

Some have found that bottle designs and fragrance formulas can be developed in North America or Europe and then produced in Asia. It is important to understand that today all aspects of fragrance development and marketing have "gone global."

Alcohol

Your finished perfume or cologne will be a mixture of fragrance oil and alcohol. Your filling house can provide the necessary alcohol and blend it with your fragrance oil but you will first be asked two questions: "What alcohol do you want?" and "How much do you want?"

The alcohol itself will be ethanol but the addition of water to the alcohol is standard practice. Water content may range from 15 percent (170 proof) to as little as 5 percent (190 proof). Today very few fragrances are blended with 100 percent pure alcohol.

The second question relates to the ratio of fragrance oil to alcohol. Will your perfume be 80 percent alcohol? 85 percent? 90 percent? The decision is up to you but you will get good advice both from your filling house and from your perfumer or fragrance creation house. Be guided by their recommendations.

(2) Finding a filling house

Ultimately the success of your project will center around the work of your filling house. All the components of your fragrance will be shipped to them. They will, in all likelihood, be the ones who will blend your fragrance oil with alcohol and they will be the ones to fill, cap, and package your bottles. They will be giving you a product that is ready to sell. Select this vendor and "team leader" with care.

How do you find a filling house that's right for you and your particular project? A personal recommendation can be helpful. Trade directories and trade shows can give you leads. A Google search can give you leads. Personal contacts might serve best, even if the contact is made through "a friend who has a friend who has a friend."

An internet search can only give you leads and help you put together a list of companies to call. Talking to someone is infinitely more effective than just sending an email or filling out contact information on a website.

Aside from an internet search for leads, there are some shortcuts you can take to get qualified referrals. Perhaps you already have a perfumer or fragrance creation house in mind to use. Call them. Talk to them. Ask them to suggest a filling house. Perhaps a bottle distributor has caught your eye. Call them, talk to someone there (be sure to get their name) and ask for their suggestions. And don't forget the competition, independent fragrance entrepreneurs you've become aware of. Call them, talk to the owners, introduce yourself and explain what you are looking for. The chances are very good that you will get the name of a qualified filling house as well as a contact name. If one competitor shuts you down, don't get discouraged. Try another ... and another. There are lots of friendly, helpful people in this business who are not intimidated by the threat of your competition.

What to look for

The first thing you want from your filling house is good communications. You want to be sure they they will listen to you, that they can understand what you want and need, and that they can help you over any rough spots that might arise during the work.

It is essential that they understand the size of your job because, if it is small, they might not want the job or, if they do agree to take it, they might delay your job while they handle their regular accounts, leaving you frustrated and uncertain when, if ever, your work will get done.

On the other hand, if your job is larger than what the filling house ordinarily handles, they may not have the automated equipment generally needed for the work and you may be charged more to have work done by hand that could, for less money, have been done by machine.

Plant tour

If at all possible and practical, have them give you a tour of their plant. See for yourself what equipment they have and what procedures have been automated and what is being done by hand.

You should be seeing work in progress and, from the labels,

you'll get some idea of what kind of accounts they have and how many people they employ.

Discussing your project

Most important of all, you'll want to discuss your project. You'll want to lay it out for them along with the names of vendors you might be thinking of using, even if you have not yet made contact with these vendors. You can expect to get some feedback. You might be discouraged from using certain vendors and encouraged to use others. Listen to the advice you are getting. It can help make your project a success.

(3) The scent — your fragrance oil

This is the "juice," as it is known in the fragrance industry, the scent, the fragrance oil that will be mixed with alcohol to produce your finished perfume or cologne. Your issues are (a) What should the scent be? (b) How much will you need? (c) Where will you get it? (d) What will it cost? and (e) Will it be "quality"?

Unless you have very strong views about exactly what your fragrance should be, and unless you are launching your fragrance into a very discerning market, acquiring your fragrance is likely to involve some compromises.

(a) What smell will it be?

When seeking a fragrance oil it helps to have some idea of what you want. If you want to become a successful fragrance entrepreneur, you should have some familiarity with the current market favorites and how your new scent will fit into the market. If your promotion will be largely based on affinity for a person, place, or event, you might go with a variation of a current best seller. This will be easy to obtain and your market will find the scent familiar, recognizable, acceptable, and welcome.

If on the other hand you are marketing to people who really know their scents and can judge both quality and originality, you will likely strive for a degree of originality that gives your prospects a small and pleasant shock, a recognition that you have created something special for them which is both new and beautiful. Achieving a successful fragrance along these lines will require a good deal of patience, taste, and the efforts of a better than average perfumer, one who understands your artistic goals and, perhaps even better than you, the commercial realities.

(b) How much will you need?

When purchasing your fragrance, your perfumer or other vendor will likely give you a quote by weight. How many pounds or kilos will you need? Your goal is to fill your bottles so you'll need whatever it will take to do that plus a little extra to allow for some waste in the filling process.

Before filling begins, your fragrance oil must be mixed with alcohol. The amount of oil you'll need will depend on both the size of your bottles and the ratio of oil to alcohol you have specified.

For example, it you are filling 10,000 one (fluid) ounce bottles, you'll need 10,000 ounces of finished fragrance (oil plus alcohol). If you decide on a ratio of 7-1/2 percent oil to 92-1/2 percent alcohol, you'll need only 750 fluid ounces of oil.

But now comes the tricky part. Fluid ounces are a liquid measurement. Your vender is quoting oil by weight. How many pounds or kilos of oil will you need to get your required 750 fluid ounces? To make this calculation you'll need to know the volume per weight of the oil, the "fluid ounces per pound" for example.

Usually you can get this information from your perfumer or fragrance oil vendor. Describe the job, the bottles to be filled and the ratio of oil to alcohol and the perfumer or vendor should be able to tell you how many pounds or kilos you'll need.

When you're working with a fragrance house to develop your scent they may do the creative work without charge provided you give them an certain minimum order for the oil once you have approved the scent they have created for you.

The formula itself — the formula developed from your ideas of what the fragrance should be — will remain the property of the perfumer or fragrance house that developed it for you. You will never see it. The closest you will get to knowing what aroma materials have gone into your fragrance will be the list of materials considered to be potential health risks that are required, in certain countries, to be disclosed on your label.

When you need more of your oil you can only go to the company that created it for you.

(c) Where will you get it?

Most of the fragrance sold by the top global fragrance and cosmetics marketers is produced for them by no more than five or six global fragrance creation giants. Do no expect them to welcome your business. You can't afford them and they can't afford you. But, if you manage to make contact with the right person at one of these giant companies, you could be given a lead to a smaller fragrance creation house or an independent perfumer who could be of assistance, either by taking on your business themselves or by guiding you to someone who will willingly work with you.

A faster and more direct way to find a qualified perfumer or small fragrance creation house is through the people at your filling house. Since they do filling, they know the companies that deliver fragrance oil to them and have a good sense of their capabilities and how appropriate they might be for your job.

When you talk to the people at your filling house and explain your project, they will almost certainly be able to guide you to an appropriate source for your fragrance oil. If they can't, either you have made a mistake in selecting them or you have described a project they don't see as being practical.

(d) What will it cost?

Every fragrance oil is different so every oil will have its own cost based on the cost of its ingredients and any extra special effort that went into its creation. A particular scent might be offered to you at different prices depending on what ingredients have been used. Often closely similar scents can be achieved through the use of different ingredients, some cheaper than others.

When you are ordering in bulk from a regular commercial source, you can expect the prices quoted to be reasonable. A large order might reduce the price per pound to some extent but your order is not likely to be large enough to achieve any real price reduction.

Ballpark figures for a fragrance might go from a low of about $20 per kilo (about 2.2 pounds) up to a bit under $100 per kilo. Although these estimates might not apply in your case, be careful if you are quoted something wildly outside this range. You are looking for a commercial source, not a vendor who typically sells a few ounces of oil at a time.

(e) Will it be "quality"?

The quality of your fragrance will depend largely on the skill of the perfumer creating it. Oddball requests from a client can sometimes insure that the fragrance will *not* be "quality," even though it conforms to the client's peculiar request.

"All natural," while perhaps desirable from a marketing point of view, will pose a number of problems for the perfumer including *cost*, as certain natural materials can be quite costly, *continuity*, as natural materials on reorder may differ in characteristics from those of the original order, and *performance*, the strength and persistence of the scent for the wonderful natural fixatives of bygone eras are mostly banned today over ethical and health concerns.

The goal of the perfumer is to produce a creation that exhibits beauty and harmony and conforms, in so far as is possible, to your specifications. In most cases if you are working with a skilled and experienced perfumer, once a general description of the desired fragrance has been agreed on, the highest quality will be achieved when you allow the perfumer to work with only a few gentle suggestions and corrections.

(4) Bottles

In most cases when selecting your bottle you will go with "stock," that is, an already existing bottle from either a manufacturer's distributor or a surplus bottle clearing house.

Production of a custom designed bottle, one that will be unique to your fragrance, requires design work, technical drawings, mold making, and typically a minimum order of 50,000 bottles or more. All of this is expensive and not generally practical for a new fragrance marketer as these costs, spread across too few bottles, make it difficult or impossible for your venture to be profitable.

Stock bottles are available in hundreds of shapes and sizes.

When decorated with your label the bottle becomes unique to your fragrance. Manufacturer's distributors are the prime source of fragrance bottles. Their offerings are cataloged and they offer continuity of supply. When you need more of the same you can generally get them.

Clearing houses sell mostly closeouts and remainders although, for a small number of very popular bottles, they may also be a distributor with a regular stock of those particular bottles. The virtue of the clearing house is that you can get "deals" on bottles, prices considerably lower than what a manufacturer's distributor would charge. The downside is that the available quantity is generally limited and next year when you need more, the same bottle may not be available. You will find too that matching a bottle from a clearing house with a proper closure, with all its necessary components — from that same clearing house — can be difficult.

Bottle descriptions

Bottles are described by size, shape and "finish." *Size* refers to the fluid capacity of the bottle. *Shape*, to its general shape category, and "finish" to the configuration of the bottle's neck. While *size* — capacity — is easily understood, and *shape* can be seen visually, *finish* is a technical specification for the bottle's neck and the closure it will require.

Finish

Perfume bottles are available in two types of finish: continuous thread and crimp.

Continuous thread

A continuous thread neck permits closure by a screw-on cap or threaded, screw-on spray pump. Continuous thread necks are ideal for hand assembly. No tools are required to screw on a cap or spray. Moreover, the closure can be removed for refilling the bottle and then reattached. If you are producing a short run — say no more than 500 or 1,000 bottles — continuous thread neck bottles are the practical choice.

There are, however, drawbacks to this choice. Sometimes a low profile look is desired and screw-on pumps rise higher above the neck of the bottle than their crimp-on cousins. This is due to the need for a certain number of threads in the glass neck to hold the spray securely. If you are closing your bottle with a cap

Bottle with continuous thread (screw-on) neck. This neck is the practical choice if you are producing just a small number of bottles of your perfume.

rather than a pump, this would not be a problem. Today there are some low profile screw-on sprays but finding one to match an available bottle can be difficult.

The second issue with thread neck bottles is that there are far fewer choices for spray pumps. You will almost certainly be selecting either gold or silver. Should you chose a cap rather than a spray, you will also find few choices in cap design and available colors may simply be black or white.

Finally, if you are producing bottles by the thousands, automatic filling machines are more commonly set up to work with crimp neck bottles and pumps.

Sprinkler neck bottles

Sprinkler neck bottles are continuous thread neck bottles with a constricted opening so that fragrance can be splashed out rather than poured. These bottles can be particularly appropriate for a men's cologne. (See an example of a sprinkler neck bottle on page 8.)

Because of the constriction in the neck, you cannot mount any kind of spray to these bottles. They can only be closed with a screw-on cap.

Orifice reducing plugs

A regular open neck screw thread bottle can be made to function as a sprinkler neck bottle by inserting a plastic orifice reducing plug in the bottle's opening. These plugs can, with some effort, be inserted manually but are intended for insertion by machine. They are available with different diameter orifices to allow more or less fragrance to be released when sprinkling.

Plastic orifice reducing plugs achieve the same effect as a sprinkler neck. The fragrance can be splashed in modest amounts. At right, with orifice reducing plug inserted, is a continuous thread neck bottle identical to the one shown on the previous page.

Thread size

Two numbers define the configuration of a continuous thread neck. The first gives the outer diameter of the neck in millimeters. The second indicates the thread configuration and height of the neck. Examples you might see could be "15/425," "18/415," or "20/401," etc.

As the purchaser of bottles your concern is simply to recognize that the finish of your caps or spray pumps must match the finish of your bottles. This is essential.

Crimp neck bottles

The neck of a crimp finish bottle will be considerably shorter than the neck of a continuous thread bottle. Ideally the lip for the pump will be quite close to the shoulder of the bottle so that a low profile will result when the spray has been attached.

Crimp neck bottles may be found with a number of different finishes such as "15 mm," "18 mm," "20 mm," etc. Each requires a spray pump sized to fit that particular finish.

Crimp neck bottles are standard for automated filling and "capping" (closing the bottle by attaching the spray or cap). Crimp-on or snap-on pumps are the only devices you can use to close these bottles. You cannot use a cap and it is unlikely you would want to use a cork!

Bottle with budget crimp-style spray pump. Because the crimp here has not been covered by a decorative collar, the crimp marks on the pump, created when the capping machine attached the spray to the bottle, are visible.

If you are producing thousands of bottles of your fragrance, you almost certainly will use a crimp neck bottle with a crimp spray pump of matching finish.

(5) Closures — caps and spray pumps

"Closure" refers to the device that will seal your bottle. This will be either a screw-on cap or a fine mist spray pump. For feminine fragrances, a spray is preferred; for certain masculine colognes, a cap on a sprinkler neck bottle can be an excellent choice.

Caps

Screw-on caps are available in sizes that will match the threads on thread neck bottles. They will share the two finish numbers of the bottle, such as "15/425," "18/415," etc. An exact match with the bottle is essential.

Caps can be custom made when the look of the bottle is particularly important. This is generally practical when you are ordering a large number of caps, 50,000 or more. If you go with a custom designed cap be one hundred percent sure that it will fit and seal your bottles properly, without the slightest trace of a leak.

In most cases if you want to use a screw-on cap you'll simply pick an existing design. Choices are limited when placing a small order but options open up as your order size increases.

Fine mist spray pumps

Spray pumps are available in three configurations: screw-on, crimp, and snap-on. Screw-on sprays go with continuous thread neck bottles and can be attached by hand or machine. Crimp style pumps go with crimp finish bottles and can only be mounted with a crimping machine. Snap-on pumps, often difficult to find, can be attached (with difficulty!) by hand or, more commonly, by machine. All pumps, regardless of style, must be of the same finish as your bottles.

Screw-on sprays

Screw-on sprays are popular for the simplicity with which they can be properly attached to bottles. You just screw them on and they make a perfect seal. Generally the actuator — the button you press to make them spray — will be covered by an overshell to prevent accidentally spraying.

Crimp style sprays

Crimp finish are the most common sprays for commercial production. Filling houses are almost certain to be able to work with them on their production lines. Aside from the need to match the finish of your spray with the finish of your bottle, crimp finish sprays can come with several other complications — or opportunities — depending on your design aesthetic and discussions with your vendor.

While screw-on and snap-on pumps come as complete units, crimp finish pumps can be purchased as individual components: the dip tube, the "engine," the actuator, collar and overshell. If shopping through a clearing house, assembling matching components can be a problem. On the other hand, if you are working toward a custom look, selecting individual components can give you the flexibility you want.

Crimp pumps can be obtained in high or low profile and the actuators (which you push to make it spray) can be obtained in various styles. Collars, the band that goes around the crimp, can be customized; dip tubes can be obtained in color or even in an "invisible" style. Guidance is needed to put these components together properly. It is, of course, possible and practical to get the complete unit in a standard form from your vendor.

This bottle, from Europe, makes use of a snap-on spray pump but only a handful of bottles were produced and the fragrance was not intended for commercial release.

Snap-on sprays

Snap-on sprays are said to be gaining in popularity but they are not yet in widespread use. Their advantage is that the spray comes as a complete unit and, when snapped onto the bottle, the actuator rises above the collar, ready for use. Snap-on pumps are said to simplify the filling process but you might have to hunt a bit to find them and then find bottles with a matching finish.

Deciding on your bottle and closure

For all the possibilities that exist in your choice of bottle and closure, in all likelihood you will be guided by the people at your filling house to select standard components that are unlikely to cause them any problems. In all likelihood you will be urged to go with either a crimp neck bottle and pump, a threaded neck bottle and screw-on pump or, if you are doing a men's cologne, a sprinkler neck bottle with a cap. Keeping it simple is a virtue.

(6) Labels

There are many ways you can label your bottles. Labels can be a simple paper label or a complex design printed on the bottle. Your decisions on labeling may involve the quantity of bottles you are producing and the overall graphic design for your package.

If you are testing with an initial distribution to just a handful of stores, you may want to fill and label only enough bottles to serve your test market. This way if your promotion goes south you can sell or reuse the empty bottles that have not been filled and labeled. Don't print directly on the bottle. If you print on the bottle and things go bad, you are left with nothing but recyclable glass.

Text

In addition to the name of your fragrance there will be requirements for certain text on your label. What this text must be will depend on the country or region where your fragrance will be sold. In all markets you will show the name of your fragrance. You will also show your company name and some abbreviated form of address.

In most markets you will also list your ingredients. In the United States this might be "Denat (or SD) Alcohol, Water, Fragrance." In Europe and elsewhere there are requirements to disclose the presence of certain specified ingredients for which health concerns have been raised. You can get a list of those that must be disclosed from your perfumer or the company that produced your fragrance.

It is a smart move to check with your filling house for how this information must be presented, how large the type must be and whether some information can be presented on your box rather than on the bottle. Both the people who are supplying your fragrance and the people at your filling house should be familiar with the relevant regulations and should be able to guide you through the labeling technicalities which, in the end, are quite simple.

Paper labels

Paper labels for your bottles are a simple, practical choice. Typically they will be self-stick, supplied on rolls. Depending on your taste and budget they can be extremely simple or wonderfully ornate. Designing your labels is a project for you, your graphic artist, and your printer.

Decorating your bottles

Alternatives to paper labels exist. Screen printing allows you to print directly on your bottles. Other options exist from companies specializing in decorating glass bottles. Etching a design onto the glass is a possibility but risky and expensive for a new fragrance by a new marketer.

The downsides of direct printing are that you'll need to imprint a fairly large number of bottles to make it cost effective. Then, once printed, the bottles are dedicated to that particular fragrance. Even the smartest marketer doesn't succeed with every launch. If you've printed directly on your bottles and your tests fail, you are stuck with bottles that can't be re-purposed for another launch or sold through a clearing house.

(7) Boxes

Boxes can be obtained in various price ranges but they will always be custom made to fit your bottles and they will always require die cutting and a certain minimum production run due to the setup costs.

The virtue of boxes is that they give you a considerable amount of freedom for graphics and you can easily decorate them in ways that you can't decorate your bottles. Even quite simple graphics can make your final package look elegant.

Talk to your filling house before ordering boxes. They are the ones who will be inserting your bottles into the boxes so you'll want to be sure that the die cut configuration you order works properly for the way the bottles will be inserted into the boxes. If you are producing just a few thousand bottles they may insert the bottles into the boxes by hand. For a larger production run, machinery may be used and in that case it is even more important that the box configuration is right for the inserting machine.

You should also carefully coordinate your box job with the

company manufacturing your boxes. In addition to selecting the right style of box, you'll need to be careful in specifying the dimensions so that your bottles will fit snugly inside the box: front to back, side to side, top to bottom. Your box manufacturer can guide you on how much clearance you'll need for proper length, width, and height.

Finally, don't forget the UPC barcode on your box or a space for a barcode sticker if you haven't yet obtained a barcode.

If your fragrance is to be sold in retail stores, your box becomes very important. If done well, the box can have a stronger influence on sales than either your fragrance itself or your bottle. Give the box design your full attention.

(8) Filling and assembly

All the components you have ordered will be shipped to your filling house which will be expecting them and will inventory them and store them until everything has been received and your slot opens up in their production schedule.

The first component you'll want delivered will be your fragrance oil since the filling house must blend it with alcohol. Typically they will be supplying the alcohol from their own inventory. Ideally the alcohol and your fragrance oil, once mixed, will be allowed to blend for about a month before bottling begins. There are some technologies that can accelerate the blending process but rushing this process and bottling too soon will take away some quality from your fragrance.

Once your fragrance has been mixed and aged and all your other components have arrived, filling and assembly can begin. Once the filling house puts your job into production, completion may take no more than a few hours. Here's what will happen.

Compressed air will be blown into each bottle to remove any dust or debris and the bottle will be placed on the conveyer belt of an automatic filling machine. The bottle will stop under the filler which will be drawing your fragrance from a barrel and filling each bottle with its correct amount of fragrance. Once the bottle is filled — it takes a few seconds — it will move on down the line to a capping machine which will put the pump in place and crimp it onto the bottle. Farther down the line an overshell will be placed on the bottle by hand or by machine. The bottle will then be boxed, by hand or by machine, and shrink wrapped if this has been specified.

Once the fragrance has been boxed and wrapped the finished bottles will be packed in master cartons, ready for pickup or shipping to you.

Depending on the size of your job and the equipment available at your particular filling house there can be a number of variations on how the work gets done but regardless of the methods and equipment used, the result will be that your fragrance is now finished and ready for you to sell.

Appendix

(1) Safety

When working with perfume, developing and bottling a fragrance yourself, take these safety warnings seriously:

Fire

A one ounce bottle of perfume on your dresser need not be treated like a bomb. But a 10 gallon drum in your house or office must be approached with caution.

- NO SMOKING

- NO OPEN FLAMES

- NO ELECTRICAL SHORT CIRCUITS THAT MIGHT SPARK

- NO MACHINERY THAT MIGHT SPARK

- NO PILOT LIGHTS SUCH AS THE ONES ON YOUR STOVE OR HOT WATER HEATER

- NO HIGH HEAT

Store bulk perfume away from these hazards. Label your containers with "flammable" labels.

Fumes

Bottle perfume only in a well ventilated space! Follow the fire hazard warnings above!

Read

Read all of the safety warning on the materials you receive. If you must re-ship perfume in bulk, alert the carrier that your shipment is flammable. Do NOT try to deceive the carrier! Your trucker or delivery service will give you the necessary paperwork and instructions.

These safety warnings are no joke

(2) Naming your fragrance

A good name for your perfume is useful. It won't make the fragrance "sell itself" but it can help make sales. It is worth taking time and putting some thought into a name for your perfume, even if you started your project with a name already in mind.

Naming a new perfume successfully involves both diligent research and creative inspiration. The creative goal is to find a name that gives your fragrance a story that is told (1) by the scent itself, (2) by the name itself, and (3) by the promotional support, large or small, you give it. When name, scent, and story are in harmony, your chances of a marketing success are improved, sometimes dramatically.

Is anyone already using your name?

You now have an issue of use. Is anyone else using your name or anything close to it? If so, your beautiful name could be a problem. Before you get too deeply committed to any particular name, search high and low for any other fragrance or fragranced product on the market that is using your name or anything close to it.

There are two practical ways to search. First do an online search. Put your fragrance name into Google and see if you get a match in something that relates to fragrance. Search under your name plus the word "perfume." Search under your name plus the word "fragrance." Search under your name plus the word "cosmetics." With luck none of the results will reveal a conflict.

Then go to the website of the United States Patent and Trademark Office and do a (free) TESS search. Here you are only looking for conflicts in the perfume category (classification IC 003). Sometimes you will find a mark whose registration has expired BUT just because the registration has expired doesn't mean that other party's rights to it have been lost. It might still be active in the marketplace and thus conflict with your name.

Don't be discouraged if your first name runs into a conflict. Try again. In many cases you'll find that *you* were the first to associate the name with a fragrance.

Trademark rights

Once you begin to market your fragrance under a name that nobody else has staked a claim to, you establish *rights* to that name, *whether you register it with the trademark office or not*. Your right is achieved by using your unique "mark" in "trade" (hence, "trademark"). Dreaming up a good name can't confer this right. You must produce your fragrance and put it on the market. This doesn't mean you must obtain mega sales before your name gains trademark rights but it does mean you must have a product and you must be offering it for sale ... to people "out there."

Keywords

One warning about naming a perfume. Increasingly we live in an online world. You, your business, and your perfume want to be found. The implications for your perfume's name are that it must be memorable enough and spellable enough so that seekers can enter it in searches — and yet it must not be so common that online searches pull up hundreds, thousands, or even tens of thousands of incorrect links with your link, if it exists at all, many, many pages down.

Trademark value

The name you give your fragrance has value aside from the fragrance itself. Once you launch your fragrance, the name on the bottle becomes a trademark and, unless someone else has already acquired legal rights to that name, by using it *you* acquire legal rights to the name. You do not have to register the name to acquire these rights. Registration can make your claim stronger but it is not necessary. What you must do it to put your fragrance on the market with the name and you should document not only the date you put your fragrance up for sale but also the date of your first sale. If at a later time someone wants to purchase your rights to the name, this information will be important.

Note that you can't just slap a name on a bottle and put it on the shelf or sell a bottle or two to your relatives. To obtain trademark rights, your efforts to sell your fragrance must be serious. That does not mean you need a large advertising budget. Your market could be eBay or your blog or Facebook. But there must be an effort to *sell* the fragrance to acquire legal rights to that name, and the name can't be in use by someone else.

Three part harmony

While there is no formula for finding a "perfect" name for your fragrance, it is helpful if your name is in harmony with both the fragrance you create for it and the promotional story you give it in your marketing. For example, if your fragrance looked black (this would be quite unusual!) it could be odd and perhaps not so helpful if you named it "Pink."

Similarly, using the name "Tiger" for your fragrance while using Egyptian pyramids for your marketing story could seem a bit odd and perhaps not so helpful.

On the other hand, if you were to name your fragrance "Pine Forest" and the fragrance had a pine scent and your advertising theme was a romantic adventure in a pine forest, you could have created a memorable harmony with each element supporting the others.

Avoid

Generally you'll want to avoid a name that is hard to remember or hard to pronounce. Also, if you sell globally, consider the impression your name will make in other cultures.

(3) Sources

The information in this book is relatively timeless but vendors, consultants, and service companies come and go. All are gone from our original project. They were listed in the original version of this book.

Instead I direct you to our website and its vendor listings that are reviewed and updated periodically. From these listings you can get leads. From these leads you can make calls and make contacts. For small projects you can order everything you need online. For larger projects the telephone is an essential.

Our vendor listings can be found here —

http://www.PerfumeProjects.com/perfume/vendors.shtml

There is a link to the vendor listings from our home page:

www.PerfumeProjects.com

Afterthoughts

This book has explained *how* to make perfume at the lowest possible cost to give you the highest possible markup whether you are selling directly to the public (as we did) or through retail stores or other distribution.

Achieving a high — or in our case, *very* high — markup is useful only if you have a marketing opportunity to *sell* your perfume (we did.) The *opportunity* always comes first.

When a ripe marketing opportunity arises, it is essential to correctly judge the *size* of this opportunity. Is it a 500 bottle opportunity? A 10,000 bottle opportunity? Or perhaps an 100,000 bottle opportunity? You must make a judgment *before* you pour money into your project. If your estimate is too far off you either produce too much inventory, dragging down or wiping out the profits from what you sold, or you produce too little inventory and miss out on profitable sales you could have made. None of us have a crystal ball to reveal the correct numbers but anyone who hopes to make money selling perfume must give serious attention to the question.

Now that you know how to produce perfume and understand that the key to profits is in the marketing opportunity, it's time to start looking for your first opportunity!